I0797280

WHAT IS
ARTIFICIAL INTELLIGENCE?

Artificial intelligence – often called "AI" – is the study of intelligent machines that can perform human-like tasks. It's a branch of computer science.

The scientists who study AI are called **AI RESEARCHERS.**

CAN AI CHANGE THE FUTURE?

DISCOVER THE SCIENCE BEHIND **ARTIFICIAL INTELLIGENCE**

(ar-tih-FIH-shul in-TEL-uh-junce)

Written by Olivia Watson

Illustrated by Denis Alonso

Words that are tricky to understand are in **bold**. Find out what they mean in the glossary.

Words that are difficult to say are in *italics*. Find out how to say them at the back of the book.

Since **computers** were invented, people have wondered whether machines could think and learn like we do. Scientists began creating special **software** called artificial intelligence, or AI.

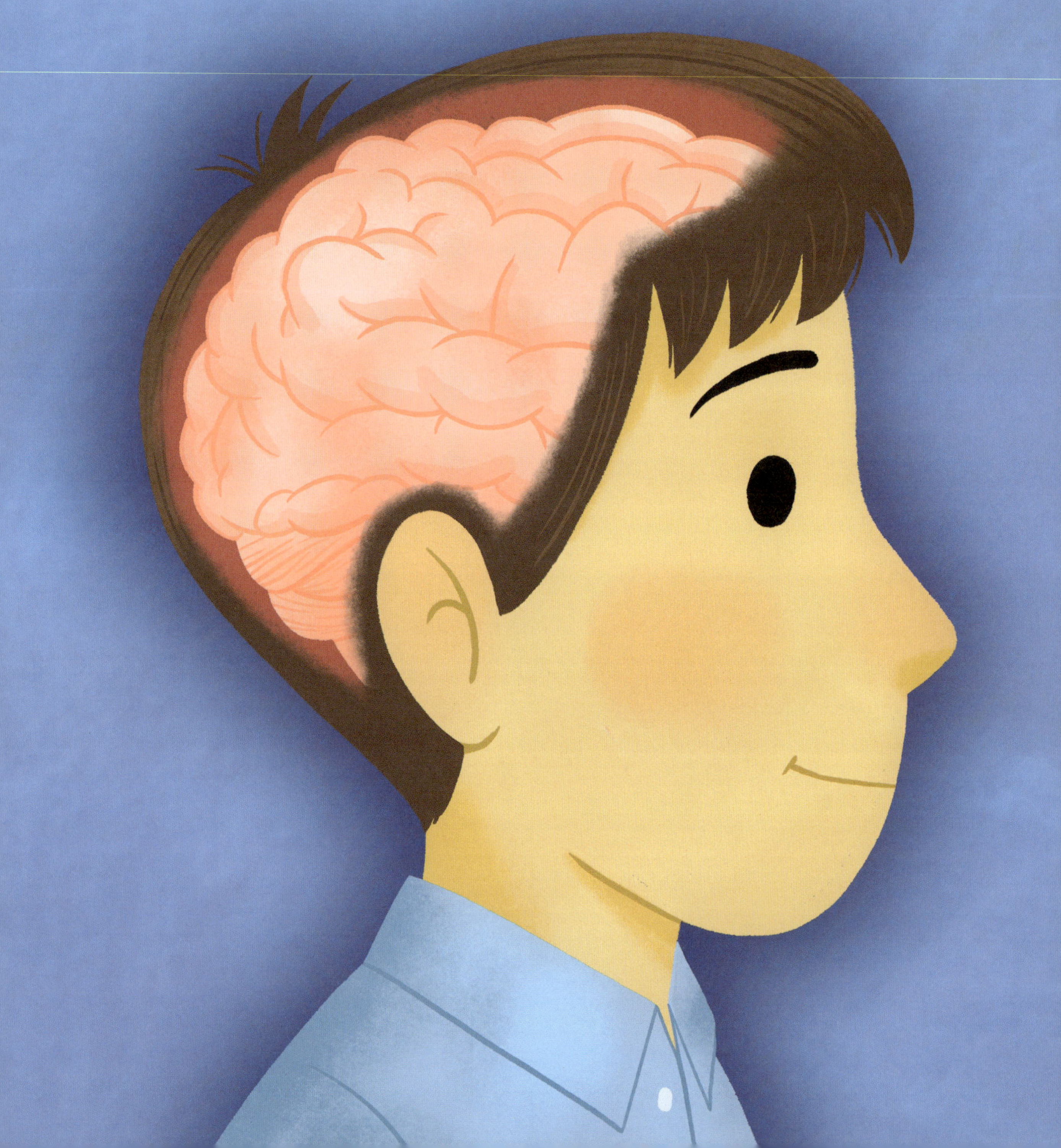

Its design copies the way human brains work, with lots of sections working together. This means AI doesn't just teach computers to follow instructions, it teaches them to notice patterns, remember experiences, and make decisions.

It took many years, but finally *AI researchers* succeeded...

One day in 1997, a **supercomputer** beat the world chess champion in a chess tournament! This was a huge victory for the science of AI. But just how did it do it?

Just like humans and animals, AI isn't amazing at doing things straight away – it needs training. Training AI is different from training a pet though! AI researchers must give AI lots of good quality information before they can expect it to be good at its job.

Now that scientists have created this clever technology, what can we do with it? AI already helps us in many ways...

AI can be **programmed** to have conversations! Because it can understand language, answer questions, and solve problems, it's already helping companies and their customers.

AI is so good at creating new things, like pictures, videos, and sounds, that it's being used to make video games more **realistic** by reacting to the player's actions!

It's not just solving problems and creating things – AI can also detect things. When teamed up with robots, AI can perform anything from cleaning tasks, like this one, to surgeries in hospitals!

In fact, AI technology could make an even bigger difference in the world of medicine. AI researchers think it can be used to look for, and **diagnose**, illnesses early. AI can process **lots of data at once!**

Finding illnesses early means people get the treatment they need sooner, and recover more quickly too. This means AI could save lives!

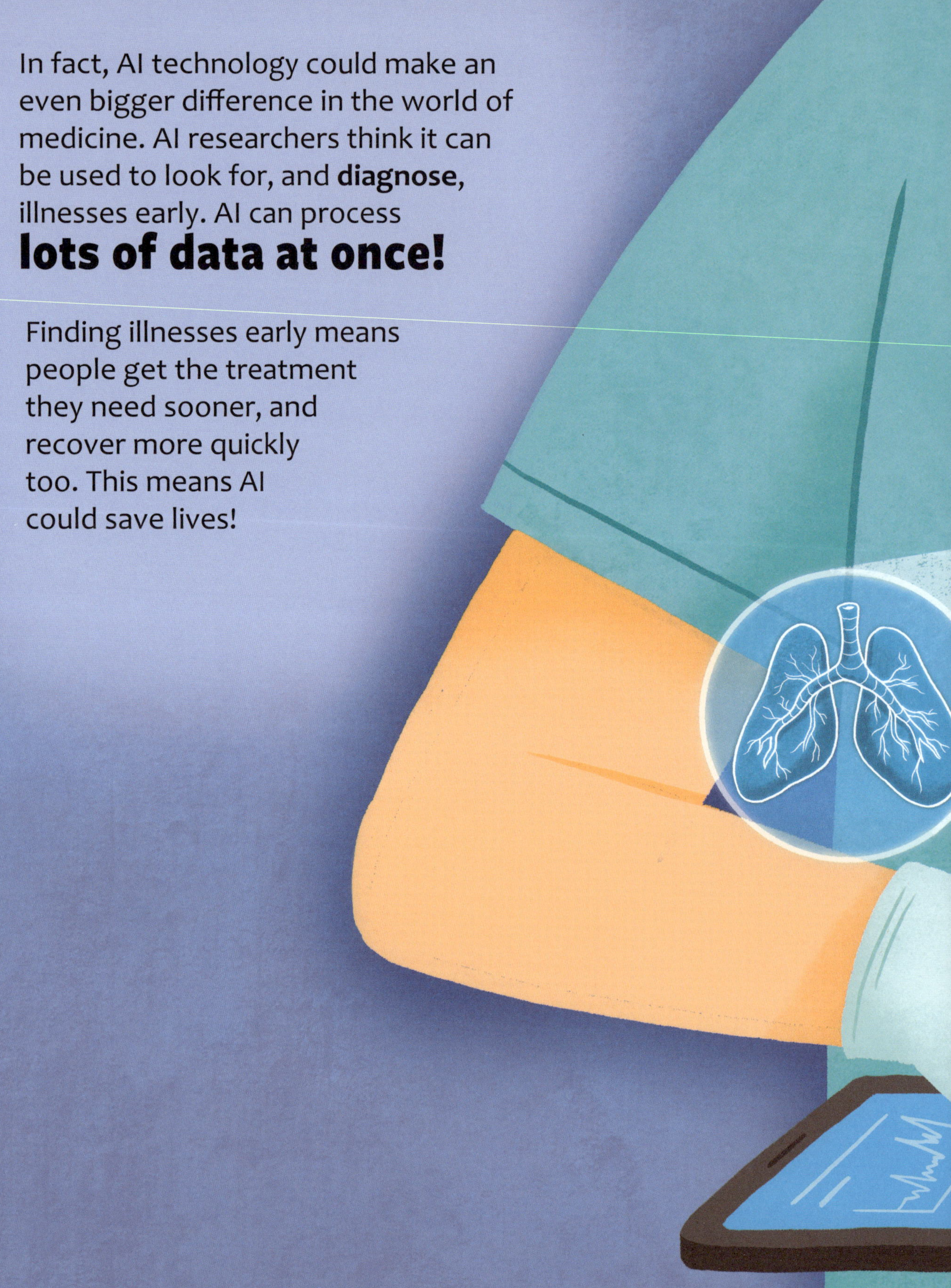

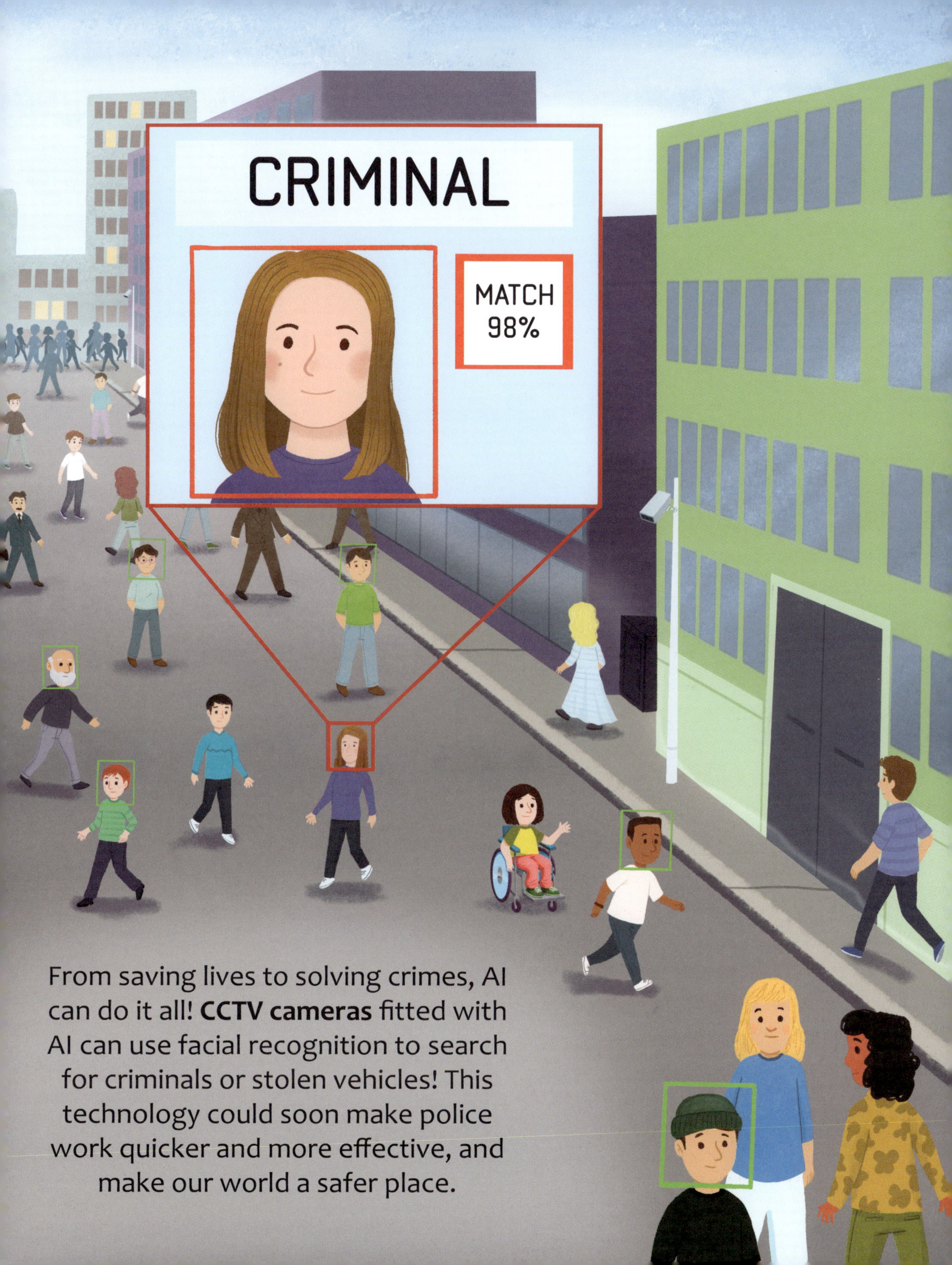

From saving lives to solving crimes, AI can do it all! **CCTV cameras** fitted with AI can use facial recognition to search for criminals or stolen vehicles! This technology could soon make police work quicker and more effective, and make our world a safer place.

Not only could AI make the world safer, it could also make our lives happier and healthier through the food we eat. Feeding everyone in the world is a challenge, but scientists think AI can help.

Machines fitted with AI could check plant health and soil quality, and alert farmers when to water, add **nutrients**, or **harvest crops.** This could help farmers increase the number of crops they grow!

Our futures start with education, so it's not surprising that AI researchers hope to help schools too! AI tutors could help by checking lesson plans, searching for answers and information, and marking homework or test results, which would give teachers more time to help students learn.

AI ASSISTANT
IQ
65%
58%
74%

DRIVIN

Some people may already be going to school in a driverless car or bus. If not, it won't be long – AI researchers around the world are creating

self-driving cars!

AI can sense traffic lights and other vehicles and, unlike humans, it won't get distracted or drive dangerously, making travel safer.

It's not just things on Earth that AI could help with. Combining AI with **robotics** could even help scientists make

discoveries about outer space!

Robots that can **navigate obstacles,** collect samples, and do experiments without being controlled by humans would completely change space exploration! These machines would allow us to learn about places that are too difficult for humans to travel to.

While AI could help with lots of things, some people worry that this technology will become too smart – even smarter than the humans that made it! Scientists call this idea “superintelligence”.

Although superintelligence could help solve problems humans have been unable to figure out, the worry is these machines may become out of our control...

Perhaps superintelligent AI would become **self-aware**, no longer wanting to follow instructions and help humans, but develop their own wants and needs, like

taking over the world!

Don't worry – superintelligence doesn't exist yet, and scientists aren't sure it ever will! Humans don't just use logic to make decisions, they use imagination and emotions too, which are things machines may never learn.

For now, scientists are programming AI to do lots of amazing, helpful things. There's no doubt that AI has the power to change the future of our world **in many positive ways!**

The Great Debate:

IS AI GOOD OR BAD?

AI can do lots of amazing things but, like everything, it has some downsides. Here are some "for and against" arguments for you to make up your mind if it's good or bad overall.

FOR: *Saving lives*

AI is already being used to help predict natural disasters like earthquakes, floods, and **hurricanes**. The earlier the predictions, the more people that can be **evacuated** and saved.

AGAINST: *Expensive*

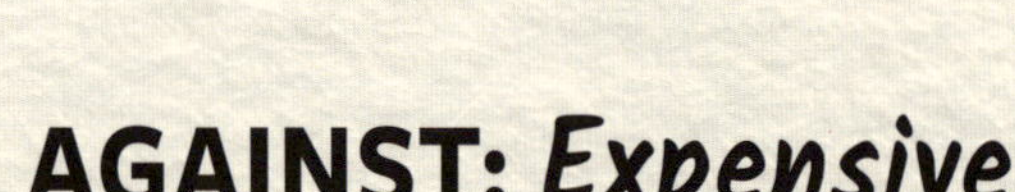

Depending on the type of AI that companies need, what the software is going to be used for, and how complicated the technology is, it can be incredibly expensive to create.

FOR: *Saving time*

Unlike humans, who need time each day to eat, sleep, and do things they enjoy, AI is available all the time. This makes it more efficient than people as it can get a lot more done.

AGAINST: *Difficult to make*

It's not easy designing and building AI systems. The most complex type of AI may take many months, or even years, to build. Once built, they need to be fed lots of data, which can take even longer.

FOR: *Less risk for humans*

AI built into robots would be able to conduct experiments and perform work that would be too dangerous for humans to do, for example working with **radioactive** materials.

These are just a few "for and against" points. How many more can you think of?

Tech-tacular

AI FACTS

There's so much to discover about the world of artificial intelligence. Did you know these fascinating facts about AI?

AI UNDERSTANDS OUR LANGUAGES, BUT ITS OWN LANGUAGE IS DIFFERENT!

Like computers, all information given to AI is written in binary code – patterns of zeros and ones – rather than words!

AI ISN'T A NEW IDEA!

The world's first robot was imagined more than 2,500 years ago in ancient Greek **mythology**. The giant robot was said to protect Princess Europa of Crete.

THE FIRST HUMANOID AI ROBOT ALREADY EXISTS!

Named Sophia, this robot was activated in 2016. It is an official **citizen** in Saudi Arabia, which has started conversations about whether it's right for robots to be considered citizens.

AI CAN DETECT HUMAN EMOTIONS!

Some AI is now able to identify human emotions by detecting facial expressions. It is accurate almost all of the time!

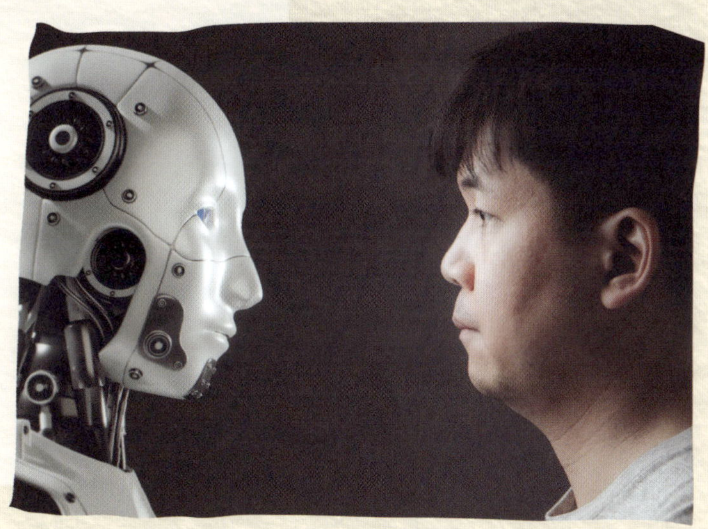

SOME AI ROBOTS CAN HEAL THEMSELVES!

Some advanced robots can notice when they've been physically damaged and repair themselves without help from anyone else.

GLOSSARY

CCTV cameras – cameras that record videos of public places to keep people safe.

Citizen – people that officially live in a specific area.

Computers – a programmable electronic device that can store, retrieve, and process data.

Crops – plants that are grown in large quantities to feed people.

Diagnose – the process of identifying an illness or injury by looking at its signs and symptoms.

Evacuated – when people are moved from an area where they are in danger to a safer place.

Harvest – to gather crops (see above) when they are ready to be eaten.

Hurricanes – a type of storm that has strong winds, rain, thunder, and lightning. They form over oceans but can move over land.

Mythology – legendary made-up stories that people once believed were real.

Navigate – move around.

Nutrients – substances or ingredients that plants and animals need to live and grow.

Obstacles – things that are in the way.

Programmed – the process of giving computers or software specific instructions to follow.

Radioactive – something that releases invisible particles that can be harmful to humans, animals, and plants.

Realistic – something that looks real even if it isn't. For example, realistic video games look like real-life videos, rather than something which has been created by an artist.

Robotics – the technology that is used to design, create, and operate robots.

Self-aware – having an understanding of one's own personality, feelings, desires, and abilities.

Software – a set of data and instructions which are used to operate a computer.

Supercomputer – a very large and powerful computer, often used for scientific calculations.

HOW DO I SAY?

AI
ay-eye

Artificial intelligence
ar-tih-FIH-shul
in-TEL-uh-junce

AI researchers
ay-eye
REE-ser-chers

THE BIG QUESTIONS ANSWERED

This is more than just a series of books; it is a complete resource. Accompanying each book is a variety of FREE material to engage curious kids with science.

www.thebigquestionsanswered.com

Use the QR code to visit the website, download free resources, and discover other books in the series.

On the website, find out incredible things about AI researchers, including what they do, some of their greatest discoveries, and the people who have made a difference in this field of science.

The material is also available for home or classroom use, supporting all the information in this book.

Teachers' & Parents' Resources
With discussion prompts and questions, extra information, and facts around key topics.

Young AI Researchers' Activity Pack
Fun activities for wannabe computer scientists, including creative writing, drawing, word searches, and much, much more.

The Big Questions Answered is published by Beetle Books. Beetle Books is an imprint of Hungry Tomato Ltd.

First published in 2025 by Hungry Tomato Ltd
F15, Old Bakery Studios, Blewetts Wharf, Malpas Road,
Truro, Cornwall, TR1 1QH, UK.

ISBN 9781835691410

A CIP catalog record for this book is available from the British Library.

With thanks to:
Editors: Holly Thornton and Millie Burdett
Senior Designer: Amy Harvey
The team at Beehive Illustration

Information in this book is up to date as of the time of writing.

Printed and bound in China.

Picture Credits:
(t = top, b = bottom, m = middle, l = left, r = right)
Wikipedia: By Unknown artist - Jastrow (2006), Public Domain, https://commons.wikimedia.org/w/index.php?curid=828070 34bl. Shutterstock: Andrey Burmakin 33tr; Anton Gvozdikov 35tr; asharkyu 33ml; Panya7 35mr; photka 32bl; Victor Ochando 34mr.